The coming of the motor lorry provided the breweries with a faster and more efficient means of local distribution to their pubs. This is Mitchells and Butlers Cape Hill Brewery yard in the 1920s: a considerable fleet is assembled for the daily delivery.

BREWING AND BREWERIES

Maurice Lovett

Shire Publications Ltd

CONTENTS

How beer is brewed 3
The history of brewing 8
The great breweries 13
Developments in the nineteenth century .. 17
The twentieth century 27
Further reading 29
Glossary 30
Places to visit 32

Published in 1996 by Shire Publications Ltd, Cromwell House, Church Street, Princes Risborough, Buckinghamshire HP27 9AA, UK. Copyright © 1981 and 1996 by Maurice Lovett. First published 1981, reprinted 1985; second edition 1996. Shire Album 72. ISBN 0 7478 0314 5. All rights reserved. No part of this publication may be reproduced or transmitted in any form or by any means, electronic or mechanical, including photocopy, recording, or any information storage and retrieval system, without permission in writing from the publishers.

Printed in Great Britain by CIT Printing Services, Press Buildings, Merlins Bridge, Haverfordwest, Dyfed SA61 1XF.

British Library Cataloguing in Publication Data. A catalogue record for this book is available from the British Library.

ACKNOWLEDGEMENTS
The publishers acknowledge the help of P. J. Ogie, Head Brewer, Whitbread London Ltd. Illustrations are acknowledged as follows: author's collection, pages 2, 3, 7 (both), 15 (bottom), 16, 19 (top), 20 (both), 21, 22 (top), 24, 26 (bottom), 27; Barnard, *Noted Breweries of Great Britain and Ireland*, pages 11, 13, 18; Bass Ltd, page 4; Bass Museum, pages 17, 19 (bottom), 25 (top); Bickerdyke, *Curiosities of Ale and Beer*, page 9; Circle Photography, pages 5, 6 (bottom); Hall and Woodhouse, page 26 (top); Michael Hardman, page 25 (bottom); Marston, Thompson and Evershed plc, page 27; Mitchells and Butlers, pages 1, 22 (bottom), 23; Tolly Cobbold Ltd, pages 6 (top), 15 (top).

COVER: *The Park Brewery, Wolverhampton, as depicted on a nineteenth-century showcard issued by the Wolverhampton and Dudley Breweries Ltd.*

BELOW: *Pubs have traditionally been centres of community activity. Today organised trips are common, but they had their beginnings when motor transport became popular: this trip is leaving the Duke of Wellington at Fenton in the Potteries in the early 1920s. Interest is such that two charabancs have been hired for the occasion. The Duke of Wellington still stands today.*

A traditional floor maltings. The soaked grain is spread out on floors and allowed to germinate, being turned at frequent intervals for consistency. Many such maltings are still in use and the most modern tower maltings still work on the same principle, although the germinating grain is now turned by machine instead of by hand.

HOW BEER IS BREWED

The traditional raw materials used in brewing beer are malt made from barley, hops, sugar, yeast and water. Cereals other than barley have often been used, but barley was the commonest; it was one of the earliest cereals to be cultivated and produced very good flavours. Sugar is used because it contributes to the colour, flavour and body of different beers.

The unique bitter flavour and aroma of beer come from resins and oils contained in the hop flowers or *cones*. Yeast also influences the flavour but its main function is to convert the sugar to alcohol and carbon dioxide. Water is often known as *liquor* to the brewer and in many breweries today the different minerals found in local supplies influence the type of beer produced.

Malt is made from barley in three stages, steeping, germinating and kilning. Malting is carried out either by independent maltsters or by the brewers. The barley is first steeped (soaked) in water and then spread on floors or put into special boxes or drums to allow germination to take place at controlled levels of temperature and humidity. When rootlets and shoots appear, the malt is dried in a kiln to stop further growth. During the germination process, important changes take place. Inside each grain there is a store of food in the form of starch and protein. The germination process releases enzymes which will change the starch into sugar. The kilning process halts the germination and thus enzyme activity. Kilning also gives malt its colour and flavour; darker beers are produced with malt which has had longer kilning.

MILLING AND MASHING

At the brewery the malt is crushed in a mill to crack the husks and becomes *grist*. This is mixed or mashed with water in a mashing machine at a carefully controlled temperature and then transferred to a *tun*. In the mash tun, the starch in the grist is converted by the enzymes to sugars. These dissolve in the water to form a sugary li-

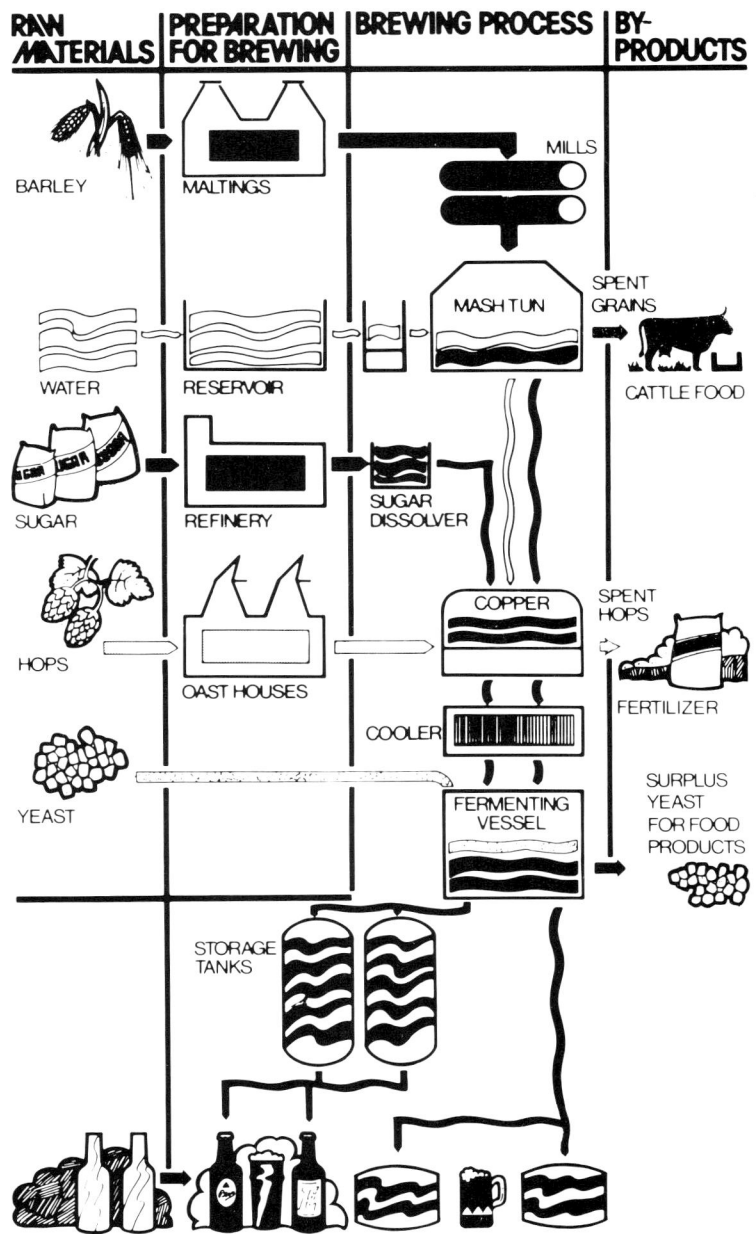

How beer is brewed.

The first step of the brewing process is mashing, where hot brewing water is mixed with malt produced from barley to give wort. The process takes place in these giant mash tuns.

quid known as *wort*. This is drawn through the perforated bottom of the mash tun down into a collecting vessel. More hot water is then sprayed on to the *mash* to remove any remaining sugar. This process is known as *sparging*.

BOILING

From the collecting vessel at the mash tun the wort is transferred to a copper, the hops are added, and the wort is boiled for one to two hours. At this stage it is possible to add some extra sugar, if necessary, for certain types of beer. The boiling process extracts the bitter flavour of the resins and oils from the hops, giving the beer its flavour, character and *nose*, stops further enzyme activity, sterilises the wort and, according to the boiling time, controls the strength of the beer.

Spent hops are removed after boiling by filtering through another perforated vessel, the *hopback*. The wort is then ready for cooling, normally by passing through heat exchangers. Many breweries use the liquor for the next mash as the coolant in the heat exchangers, thereby saving energy.

FERMENTATION

The cooled wort is now run into a fermenting vessel and excise officers gauge the quantity and specific gravity to calculate the duty to be paid by the brewery. The yeast appropriate to the type of beer is then *pitched* or added to the vessel at a carefully controlled temperature. It feeds on the sugars in the wort, producing alcohol and carbon dioxide naturally in the process. Surplus yeast collects at the top of the fermenting vessel, in a *head*, and is skimmed off the surface; it multiplies about five times during the fermentation process. The surplus yeast is used by food processors. The fermentation process normally takes from four to eight days, after which the beer is run off to mature in conditioning tanks, for keg or bottled beers, or into a tank for *racking* into casks, for

ABOVE: *The wort is boiled in giant coppers, at which stage the hops are added.*

BELOW: *Most traditional English beers are fermented in open vessels known as 'squares'.*

ABOVE: *Modern racking (cask-filling) lines are reminiscent of their predecessors; casks are rolled forward from a floor-mounted chain conveyor and the fillers work on a back pressure system. The basket holds the wooden bungs.*
RIGHT: *A modern cask-filling line: the operator is 'dry hopping' each cask, dropping compressed dried hops through the bung hole as the cask is filled.*

draught beer.

As the beer is run into the casks many breweries add dried hops to give extra flavour and aroma, and immediately before the beer leaves the brewery a small quantity of isinglass finings is added. These will cause the yeast still present to settle on the bottom of the cask, leaving the beer bright and clear. The beer then comes into condition naturally in the pub cellar.

THE HISTORY OF BREWING

Beer as we know it today is a much older beverage than people imagine. Brewing, using malted cereals such as barley, was common in the Bible lands, and the ancient Egyptians even had a hieroglyph for brewer. This early brewing was crude but well established: the techniques of malting and fermentation were understood, although both were almost certainly originally discovered by accident.

The earliest brewers would have produced their malt by first soaking barley in water, then burying it in the ground to allow it to germinate. The malted barley was then dried, crushed in a mortar with a pestle and then baked into large flat cakes. When brewing took place, these loaves were soaked in warm water and pressed through a sieve or basket in a simple mashing process. The resulting wort would be fermented in large earthenware vessels, probably using yeast saved from a previous batch of beer, or even the natural yeast contamination of a cracked vessel. Variations in technique were common: in some areas the loaves were mashed into a thick porridge and, after being allowed to stand for a time, were sprayed with more water – a technique known today as sparging, which is still common in many breweries.

Beer produced by these methods would have been unusual by modern standards. Although there was often some form of filtration during the mashing process, most drinkers were unconcerned about cloudy beer. One type was so thick with unfiltered barley husks that it had to be drunk through a hollow reed used like a modern drinking straw, although this was turned to advantage and it was a sociable custom to share a drink with one's friends! These early beers were drunk at all stages of fermentation, so that both weak and strong varieties were easily obtainable. Flavourings were also added, using various herbs or plants.

Babylon and Egypt were the most prominent brewing countries in ancient times; evidence discovered by archaeologists suggests that techniques became more and more sophisticated and the brewer achieved an important status in society. This may well have been associated with the supply of beer to the temples.

The introduction of beer into Britain has often been credited to the Romans but it was certainly known and drunk in Britain long before that, with native grown cereals being used as a basis. The British climate is ideally suited to growing cereals, and the farmers of the neolithic period almost certainly produced a form of beer. The technique of brewing had probably travelled overland from the Middle East and may well have arrived at the same time as the barley which was its essential ingredient.

The Romans found beer making well established when they arrived in Britain, although it was not to their taste. They introduced wine, both for their own consumption and for Britons who became Romanised, while the common people continued to brew and drink their own beer. Roman writers looked down on this native drink. Pliny the Elder wrote: 'western nations intoxicate themselves by means of moistened grain.' And the Emperor Julian said of British beer: 'Who made you and from what, by the true Bacchus, I know not. He smells of nectar, but you smell of goat.' But the local climate, except in the south of the country, was hardly suitable for viticulture, and after the Romans left in the fifth century the British continued to make their beer.

Brewing was carried out at home and on the farms, normally by the women of the household. There was no brewing in breweries as we would understand them today and the beer available in alehouses would have been produced on the premises. A substantial amount of brewing was also carried out in the monasteries and much of the beer that was produced was supplied to travellers or pilgrims who used the accommodation provided by the monks. As the number of pilgrimages to shrines and other holy places increased, hostels, which were the forerunners of inns, were built by the monasteries. The monks themselves also consumed considerable quantities of beer. Burton Abbey, founded by Wulfric Spot, Earl of Mercia, in 1004, allowed its monks 'one gallon of strong ale often supplemented by one gallon of weak ale' daily.

The Normans reintroduced wine in large quantities to Britain but by then beer was the staple drink of the people. As a period of stability came to Britain, travellers began to demand better facilities 'on the road'. The monasteries provided establishments organised on a commercial basis, and inns which provided food, drink, stabling and a bed for the night became more common. Beer was drunk widely by men, women and children, chiefly because it was safe to drink, unlike water and milk, the other two main beverages. The principles of brewing – heating water before mashing, boiling the wort and fermentation which produced alcohol – ensured the liquid was sterile, and more important, remained so.

As brewing became more organised, it attracted the attention of the tax collectors: the main fairs of the Middle Ages placed a toll on ale sold from barrels at the fairgrounds and other special local taxes were often due. These were usually paid to the lord of the manor or to the abbot of the monastery. Henry II introduced the first national tax on beer in 1188 to raise money for the Crusades – the so called 'Saladin Tithe' – and since then it has always been taxed in one form or another.

Brewing was becoming an important trade and the brewers of the inns and alehouses were developing a wider range of beers to appeal to their customers.

By the middle of the thirteenth century, while brewing was carried on in the taverns, alehouses and inns of towns and villages, it remained an important part of the woman's work in the home and on the farms: much of the general brewing was also carried out by women – the 'ale wives' or 'brewsters.'

The price of beer and bread was fixed by Henry III in his Assize of Bread and Ale in 1267, and the price of ale was constant throughout this period at 1d, 1½d and 2d per gallon, depending on the strength. Also during this period more controls were introduced on the brewers, concerned with maintaining acceptable standards of quality and price: ale conners were employed to inspect and test the quality of ale on sale.

The Assize of Bread and Ale left a loophole for the unscrupulous brewer; it

The interior of a fifteenth-century brewhouse. Equipment is simple, with much use being made of coopered vessels. The brewing principles are much the same today.

fixed the prices of both these essentials in direct relation to the price of the grain needed. Naturally, the brewing process allows ales of different strengths to be produced, and this was a temptation to produce the weakest ale possible at the highest price without getting caught! The ale conners or ale tasters were appointed, usually by the local authorities, to test (by drinking) the quality of the beer sold and the measures being used. Much of their work was concerned directly with brewing: in many areas, brewers were not allowed to sell their ale until it had been given official approval by the ale conners. One legend which has grown up around their work is that they were supposed always to wear leather breeches when performing their tasks so that they could test the stickiness of the ale by sitting on a drop. It is a matter of argument whether good adhesion (plenty of unfermented sugar in the ale) indicated a better quality than poor adhesion!

It was during this period that the brewery with the world's longest record of

ABOVE: *Hops were introduced to Britain from the Low Countries, both as a flavouring and to help beer to keep. The 'bines' grow up wires on stakes and are cut down before the hops are picked. At this modern hop yard in Herefordshire, looking out over the tops of the bines, the conical roofed oasthouses, where the picked hops are dried, can be seen.*
RIGHT: *The interior of a common brewer's establishment in the early eighteenth century. Equipment is simple: extensive use is made of timber and wherever possible gravity is made to do the work. The vessels shown are: LB, liquor back; MT, mash tun; UB, underback; CB, copper back; C, copper; JB, hop or jack back; B, back or cooler; Sqr, square or working tun. The basic equipment shown would not have changed from the seventeenth-century brewhouse, except in size.*

brewing was founded. In 1340 Queen's College, Oxford, was established and, in the practice of the time, a brewer was appointed. Brewing continued for six hundred years, only ceasing at the beginning of the Second World War.

Ale was brewed without the use of hops. Beer, brewed in the Low Countries, was first imported into England and sold in London in the early fifteenth century. On the continent hops had been used to flavour beer for many years, notably in central Europe, and gradually replaced other plants because of their special ability to prolong the life of the beer. There was, however, considerable resistance to the use of hops in brewing in Britain: English ale was traditionally very strong and very sweet, which helped it to keep well. Hops achieved much the same effect in beer, but without so much strength or sweetness, introducing a new and subtler taste to the customer.

Most of the early brewers of beer came from the Low Countries and settled in London, which provided them with a huge and constantly growing market for their wares and ideal country nearby, in Kent, in which to grow hops. The brewers of ale and the brewers of beer kept themselves very much to themselves for many years, forming separate guilds and even using different sizes of cask for their products. At this time the different sizes of cask were:

	for beer gallons	litres	for ale gallons	litres
Pin	4½	20.457	4	18.184
Firkin	9	40.914	8	36.368
Kilderkin	18	81.828	16	72.736
Barrel	36	163.656	32	145.472

Cask sizes for beer are the same today, with the largest, a hogshead, holding 54 gallons (245.484 litres).

The hop was treated with great suspicion by the ale brewers and many authorities. This may well have been because of a desire to maintain the traditions of ale brewing and also suspicion of the brewers of beer, most of whom were foreigners. Even Henry VIII's own brewer was prevented from using hops as late as 1531, but gradually hops were introduced into ale and today ale and beer are indistinguishable.

During the fifteenth and sixteenth centuries the demand for ale and beer steadily increased and breweries began to produce on a larger scale, although most were still attached to an individual tavern or alehouse. The day of the 'common brewer' was still to come.

Some idea of the scale of the brewing and consumption of ale and beer can be gained from the fact that a national licensing system for alehouses was introduced by Henry VII in 1495. This was an early attempt to give justices of the peace power to take sureties for good behaviour from keepers of alehouses.

English beer was exported, too. In 1492 John Marchant, a Fleming, shipped 50 tons of 'Beere' from his newly established Red Lion Brewery in London. This brewery was later operated by Hoare and Company and was eventually taken over by another of the great London brewers, Charringtons. By the middle of the sixteenth century, there were twenty-six common brewers in London and a host of innkeeper brewers who produced sufficient for their own needs on their own premises. Elsewhere in Britain the same pattern would have been repeated on a smaller scale, although in country districts the production of malt was only carried out in the winter months, because of the difficulties of controlling the temperatures during the process.

The basic raw materials of the sixteenth-century brewer were malt, unmalted oats or wheat, hops or some other plant for flavouring, water and yeast. The working of the yeast and the whole fermentation process was a considerable mystery: it was not until the nineteenth century that brewing science had advanced sufficiently for the process to be fully understood. However, brewers would have appreciated the need to retain some of their yeast for repitching into the next brew, and brewers were expected to supply other brewers or even bakers with yeast on demand.

In 1577 a census of all taverns, inns and alehouses was taken. The total for England and Wales came to 19,759. With a total population of 3,700,000, that gave a figure of one to every 187 people. (Today, the figure is about one to every 650 people.) Once again the purpose of the survey was to raise money by taxation, this time to help pay for the cost of repairing Dover Harbour.

Although by the end of the sixteenth century puritanism was beginning to foster a reaction against drinking in all its forms, the brewers now represented a useful — even essential — form of revenue and moves against them were limited by financial considerations. During the Civil War taxes were imposed by both the King and Parliament to raise money to pay for the war. The beer duty was 2s on beer with a pre-duty cost per barrel of 6s and 6d on beer with a cost less than 6s. Hops were also taxed. It was said that these taxes would only be for the duration of the war, but like so many promises by government this was not to be. Oliver Cromwell, who was the son of a brewster, freed the domestic brewer from paying duty, but even after the Restoration excise duty remained, despite the common brewers' efforts to have it removed.

In spite of increased taxation, the seventeenth century saw the establishment of many of the famous brewing companies whose names are still well known today. Trumans, Halsey (later to become Barclay Perkins), Reids, Guinness (then operated by Giles Mee) and Calverts all came into being during this period. By the end of the seventeenth century the pattern of brewing was changing. Innkeeper and alehouse brewers were gradually disappearing to be replaced by the common brewers.

Meux's Horse Shoe Brewery, on the corner of Tottenham Court Road and Oxford Street, London, became famous in 1814 when the great vat of porter burst. Many houses were swept away in the flood and eight people drowned. Many others drank from the liquid flowing along the streets and became remarkably intoxicated! The vat was 22 feet high (6.7 m) and held 3,555 barrels of beer, over a million pints.

THE GREAT BREWERIES

In the eighteenth century commercial considerations brought about a rapid growth in the size of the breweries and in their methods of operation. An increasing population demanded ale and beer in greater quantities; it was produced in breweries whose names are still known today and which were founded after 1700. Allsop, Bass, Worthington, Whitbread, Meux, Guinness, Charrington, the Leith brewery of William Younger and Simonds of Reading all came into being during this period. Most had small beginnings but, because of entrepreneurial flair and the rapid growth of their market, grew into very large companies. Interestingly, a substantial number of local breweries also founded in the eighteenth century still survive, although most remained local, supplying their beers over a relatively small area. Brains of Cardiff, the Belhaven Brewery, Cobbolds, Hartleys of Ulverston, Greenall Whitley, Buckleys of Llanelli, Randalls of St Peter Port, Guernsey, Hall and Woodhouse of Anstey, Dorset, Boddingtons in Manchester and the Workington Brewery are just a few examples.

In an attempt to reduce the sale of gin, ale and beer were promoted as the drinks for the common man which also supported the farmer. The adverse effects of gin, particularly in London, were little short of disastrous. 'Drunk for a penny, dead drunk for two pence' was a catchphrase with meaning. Ale and beer used materials from English farms — barley and hops — and there was much encouragement to drink beer rather than gin. Two contemporary poems illustrate this.

Gin
Gin! cursed fiend with fury fraught,
Makes human race a prey,
It enters by a deadly draught,
And steals our life away.

Beer
Beer! happy produce of our isle
Can sinewy strength impart,
And wearied with fatigue and toil
Can cheer each manly heart.

In 1736 the Gin Act was passed, basically to force those selling gin to take out a licence and to impose a heavy duty on the distillers and thus restrict the number of gin shops. The act was largely ineffective: by 1738 there had been over eleven thousand prosecutions.

The strength of feeling about beer and gin was reflected in two of Hogarth's engravings, 'Beer Street' and 'Gin Lane'. In the first there is an air of prosperity; everybody thrives, with one notable exception, the pawnbroker, whose premises are in ruin. 'Gin Lane' is the complete opposite. There is squalor, degradation and poverty and only the pawnbroker flourishes. Hogarth was a contemporary commentator on the social scene and it is likely that his engravings give a true picture. But ale and beer thrived. More and more breweries were established and the variety of products available increased rapidly. Beer was part of the everyday diet of every man, woman and child in Britain, representing an important part of their daily food intake.

Brewing science and technology advanced as breweries expanded. The invention of the steam engine provided the brewer with a means of mechanising his processes, replacing the power of men and horses, which had been his basic sources of energy. Breweries quickly became major users of steam power and, with this new technology, significant changes began to take place in the traditional equipment. Metal vessels gradually replaced wooden ones, steam pumps were introduced (many of the operations of brewing involved moving large quantities of liquids from one vessel to another), mechanical rakes, screw conveyors and mechanical mashing machines were installed, and a new skill came into being – that of the brewing engineer.

But discovering what went on during the malting and brewing processes attracted the most interest. A significant development was the reinvention by Dicas, in 1780, of the hydrometer — an instrument which provided an accurate method of determining the amount of sugar in the wort. Brewers had been experimenting with home-made instruments for some years: Baverstock of Alton was using one as early as 1769.

It was during this period, too, that Fahrenheit's thermometer came into common use in the larger breweries. Much of the success or failure of particular brews depends upon control of temperatures during the process. Here, at last, was a means by which the brewer could obtain and, more importantly, record accurate information.

Patents came thick and fast. The processes of malting, mashing and boiling, the analysis of hop extract and the methods of drying yeast all came under scrutiny during the half century from 1750 onwards. Publications began to appear which explained some of the mysteries of brewing: writers began to record some of the processes in detail. Before 1750, little is known with any accuracy of the processes; the first serious works of real value were the editions of *The London and Country Brewer*, published anonymously between 1734 and 1759. Michael Combrune in his *Essay on Brewing* and *Theory and Practice of Brewing*, both published in the 1760s, also covered much technical theory and explained the use of the thermometer.

It was probably experimentation by brewers, based on the new publications available, that advanced scientific knowledge during the later years of the eighteenth century.

Isinglass, the substance still used by brewers today for clearing the sediment from the casks in the pub cellar, was first mentioned in 1695 by James Lightbody in *Every Man His Own Gauger*. By the middle of the eighteenth century it was in common use and had considerably shortened the time between brewing and drinking beer. *The London and Country Brewer* places considerable emphasis on the clarity of beer, showing that interest was growing in the appearance (and therefore the quality) of the product.

But what were eighteenth-century ale and beer like? During the early years most were brown or black in colour and very strong, although 'small' beer (a weaker version) was popular with women and children and for drinking at breakfast. The two most significant products developed during this period, however, were porter and East India pale ale or IPA.

Porter was invented almost by accident. It was common practice amongst the working classes in London to drink 'three threads', a mixture of three types of beer. An enterprising brewer — some claim that it was Ralph Harwood of the Bell Brewery in Shoreditch – combined the three beers into one and called it 'Entire Butt' or 'Harwood's Entire'. Its popularity with the porters of the great London markets led inevitably to its renaming. It was dark, almost black in colour, strong and well hopped and, after brewing, was stored for long periods in huge vats before sale. A Frenchman wrote in 1726: 'The greater quantity of this beer is consumed by the working classes. It is a thick and strong beverage, and the effect it produces if drunk in excess is the same as that of wine; this porter costs 3d the pot.'

East India pale ale was totally different. It was probably developed originally by Hodgson at the Bow Brewery in London.

ABOVE: *An early open copper still in use today. This 1723 vessel was first installed at the Harwich brewery by the Cobbold family. When they moved to Ipswich, the brewery's present home, the copper went too and is now used as a sugar dissolver.*

BELOW: *Porter became the popular drink in the cities during the nineteenth century, with production centred in London. Porter needs a long time to mature and this was done in huge vats, which were often built into brewery cellars or even excavated and lined with tiles. Those illustrated are wooden and were made by the brewery's coopers.*

Light, sparkling and pleasantly bitter, it was an immediate success both in Britain, particularly with the middle and upper classes, and overseas, where its name indicated the main area of sale – to the British in India. The earliest trade was through the East India Company, but by the 1820s the brewers of Burton upon Trent had taken up the beer. Burton's natural well water is rich in the mineral salts essential to this variety and the town's brewers established a considerable reputation for the finest pale ales. They had long been renowned for the quality of their ales; the national beer export figures of 1800, with no fewer than ninety thousand barrels being exported, were largely due to Burton's brewers shipping beer down the river Trent to the ports of Gainsborough and Hull for the Baltic trade.

Beer from Burton was also well known in London: it was sold at the Dagger in Holborn, at the Peacock in Grays Inn Lane and in Vauxhall Gardens, the capital's notorious pleasure park. Joseph Addison wrote in *The Spectator* in 1712: 'we concluded our walk with a glass of Burton ale.'

By the end of the eighteenth century brewing had increased in scale and importance. Brewers had become men of social stature, with interests in politics, a range of associated businesses and even banking. This trend was particularly evident in London; technical advances and the sheer size of their market had allowed them to outstrip their poorer country cousins, whose breweries largely remained much as they had been a century before.

Beer was exported in large quantities even during the eighteenth century. Beer was stowed in the holds of sailing ships and often taken out as ballast on ships collecting other cargoes. The wooden battens across the ends of the casks served a dual purpose: they helped to prevent damage but also stopped the crew from stealing the contents.

Many breweries were built to take advantage of the railway network. This ale dock at Worthington's Burton brewery was purpose-designed to allow easy loading and handling straight on to a railway company's wagons.

DEVELOPMENTS IN THE NINETEENTH CENTURY

The early years of the nineteenth century gave the brewers the opportunity to expand, in both the size and complexity of their breweries, and to continue to develop the scientific side of the business. Steam remained the key source of power until well into the twentieth century, but surprisingly there were still considerable gaps in scientific knowledge, even in the largest breweries, on questions such as yeast behaviour and the processes taking place in fermentation. Indeed, it was not until Pasteur visited England and Whitbreads brewery in 1871 and demonstrated the use of the microscope to examine yeast that brewers began to understand the micro-organisms that could spoil beer.

However, in other areas the advances were the most rapid that had ever been made. Technical developments were significant: cast iron mash tuns, combined with mechanical mashing or mash-mixing machines, improved and simplified the process. Steel's Mashing Machine, patented in 1853, is still an important method of mash mixing in use in breweries today. Sparging or spraying the mash with hot water came into common use; this ensures the complete use of all the goodness in the malt and replaced the earlier (and inferior) system of mashing several times over using the same malt.

But it was with temperature control that the greatest advances were made. The

The classic layout of a nineteenth-century brewery, in this case Stansfeld's Swan Brewery at Fulham in London, shown in 1890. The building at the centre with the flag is the 'Tower', which contained the main brewing equipment: liquor tanks, grist mill, grist case or hopper, mash tun, underback or wort receiver. Using this system, it was possible to contain all brewing operations on a compact site. Some of the smaller breweries even had a maltings for producing their own malted barley within the same area.

brewing process demands heating and cooling of the liquids and systems which speeded up the process were invaluable to the brewer. The nineteenth century saw developments in cooling and refrigeration whose basic principles are still in use today, notably with the heat exchanger, a piece of equipment which pipes cold water counter-current to the hot wort, thus absorbing the heat.

Increasing mechanisation required a wide range of engineering skills. A nineteenth-century brewer said: 'It is the duty of the Chief Engineer to design, erect and keep in order all buildings, engines and machinery, boilers, water arrangements, roads, drains, railways, gas, workshops, lighting and fire brigade.'

In spite of mechanisation, brewing was still remarkably labour-intensive, particularly in the 'support' operations — malting, engineering, bottling, coopering and distribution. Much of the work was also seasonal, with workers being employed at the busiest periods and laid off when demand slackened. However, there were some enlightened employers, providing what were, for the nineteenth century, extremely good conditions. The political and religious convictions of many brewers (Quakers were prominent amongst the owners of several London breweries) no doubt benefited their workers.

The larger companies, although expecting long hours of hard physical labour from their men, did provide some bonuses, usually in the form of specially brewed weak beer, available in considerable quantities to those engaged in the most demanding work. Others provided savings clubs, Bible readers, canteens, inexpensive tied housing close to the brewery (the owners and managers also lived 'over the shop') and, in one case, a coffin and an appropriate number of followers if an employee died in the service of the company. Many brewers also provided their workers with clothing and tools, often made on the premises: 'there is a tailor's shop for making flannel clothes and coats for Watchmen, a clogger's shop...' The lot of the brewery worker in the nineteenth century was usually far superior to that of his brother in other industries.

Employment in the industry varied widely. The small country breweries still employed only a few men: towards the end of the nineteenth century one is said to have employed 'thirty men who brew ten thousand barrels of beer per annum'. On the other hand, the large companies were considerable employers of labour: one major brewery in the 1880s employed over 2,700 men, who brewed almost one million

ABOVE: *Nineteenth-century mechanisation at its most sophisticated: the magnificent array of mash tuns at the Guinness brewery in Dublin.*
BELOW: *The cask-washing shed in a nineteenth-century brewery. It is vitally important to clean all containers thoroughly, as any micro-organism can quickly spoil the beer. Wooden casks can cause problems; it is easier to clean a metal cask thoroughly. Most breweries still employ 'smellers', whose sensitive noses, applied to the bung hole of a cask, can quickly detect any trace of sourness.*

ABOVE: *A large mechanised racking (cask-filling) line in a nineteenth-century brewery — in this case Guinness in Dublin. The racking 'engines' have a number of heads, so that several casks can be filled simultaneously.*
BELOW: *Bottling about 1900. Beer is being drawn from the large casks at the right of the picture. The machinery is of the simplest kind, with most operations being done by hand. Filling is being carried out on the right, on a siphon filling machine, which produced a form of automatic operation.*

Before the days of the metal cask, breweries employed large numbers of coopers. In the smaller establishments casks were made entirely by hand, but the larger breweries tried to mechanise their operations as much as possible. Substantial use was made of steam for power and this picture shows part of the Bass steam cooperage in Burton upon Trent. Today, metal casks have almost totally replaced the wooden coopered cask.

barrels of beer in a year — and that figure would not include the hundreds of clerks, warehousemen, labourers and drivers at the many depots and branches throughout Britain. Brewing was a major industry making a considerable contribution to the wealth of the country.

The use of casual labour reached its most sophisticated level in Burton upon Trent in the late nineteenth century. Bass, needing large numbers of workers for the winter malting season, brought in hundreds of East Anglian farmworkers, known locally as 'Norkies.' These men would be dismissed by the farmers in the autumn and re-employed in the spring: the Burton trip, with a single rail ticket and approximately four times their farming pay, was naturally much sought after in spite of the intensive work.

More brewers began to produce their own malt, developed more sophisticated methods and obtained a better extract. During the nineteenth century brewers paid duty on malt, on beer and even on the hops they used, but in 1830 the Duke of Wellington's Beerhouse Act abolished duty on beer and made it possible for any householder who could produce two guineas to be granted a licence to sell beer and cider.

There were a number of reasons for the Act. It was virtually impossible to gauge the strength of the fermented beer, so fraud was common; consumption of spirits, notably gin, was still increasing (it had doubled between 1807 and 1827 and persisted throughout the century) and the authorities were keen to reduce this trade; private breweries did not pay duty, so it was the customers of the common brewers, the working classes, who suffered from the tax; and tea, which had to be imported and was not taxed, was also gaining a hold on the market. However, the tax on malt and hops remained, the former until 1880 and the latter until 1862. The Beerhouse Act was modified many times over the years and remained substantially in force until 1869, when the responsibility for issuing licences passed to the magistrates.

Beer duty in the form we know it today was introduced in 1880, when the use of the hydrometer (or saccharometer) made accurate measurement of the specific gravity of wort possible and allowed the

ABOVE: *During the world wars many of the traditional men's jobs, even the heaviest, were done by women. These workers are in a maltings, but women took over ale loading, racking (filling casks) and even operated signal boxes where breweries had their own railway systems.*

BELOW: *Breweries were quick in time of war to show that they were doing their patriotic duty. This picture shows food production (grain and potatoes) in grounds adjoining the Cape Hill Brewery in Birmingham in 1917.*

The risk of fire and the shortcomings of the national fire services persuaded many breweries to establish their own fire brigades. This photograph shows the horse-drawn appliance of a brewery brigade. The men became so proficient that, as well as dealing with brewery fires, they were often the first on the scene at fires elsewhere.

government to introduce a tax on this wort gravity. Because the tax was based on the potential for fermentation, beer tax was for the first time ever related to its alcoholic strength.

By the middle of the nineteenth century brewing in Britain had divided into a number of distinct areas and had begun to establish the form of the present-day industry, with concentration of types of beer in certain areas. Local breweries remained small and specialised in the working man's drink, a brown ale similar to today's mild. Other beers would be produced to satisfy specific local demand, but these brewers would also offer the specialised beers (pale ales, for example) of the major breweries.

London brewers, because of their vast market, grew much larger and offered a wide range of beers, but with porter as their principal product. Many also opened breweries in Burton upon Trent or established links with existing brewers in Burton, to supply the unique pale ales and bitter beers of the town.

Regional brewers based on the growing cities tended to produce a wider range of ales and beers than their smaller competitors, but all their products were strictly tailored to suit local taste. Volume was what they were after, not variety.

Burton upon Trent became the pale ale capital. Other beers were brewed, but it was for its pale or amber beers that this unassuming Staffordshire town became famous and its brewers among the largest in the world.

Scotland's beers were (and in most cases still are) different, being darker in colour and sweeter than beers for the English palate.

In Ireland most brewing was concentrated on Dublin, whose specialities were at first porter and then stout. The nineteenth century also saw the beginning of the great export markets for stout, led by the Dublin brewers, Guinness.

In spite of the explosion in demand for beer, the numbers of breweries declined rapidly as the nineteenth century

Economic transport has always been crucial to the breweries and from early times most breweries maintained large numbers of heavy horses and horse-drawn vehicles. The brewer's dray developed into particular forms in different areas and an expert can identify the origins of specific vehicles by their design. This is a 'floater', a low two-wheeled cart pulled by a single horse and originating in Burton upon Trent.

progressed. In 1840 there were almost fifty thousand brewers in the United Kingdom, including small innkeeper brewers. By 1880 there were less than half that number, and by 1900 the number had declined to just over three thousand.

As numbers declined, breweries became bigger and began to buy public houses to guarantee sales of the beers they brewed. In this way the 'tied' house system, which brought about the development of the English pub as we know it today, was established. The system was born late in the eighteenth century. Brewers began to realise that their methods of distributing and selling beer were at variance with the new efficiency prevailing in their breweries. They needed guaranteed sales and saw that owning a percentage of the outlets was one way of ensuring this. Already they acquired public houses from time to time for non-payment of debts. Thus the brewers became the publicans.

It has been estimated that by 1820 more than half the public houses in London were directly controlled by brewers, and as the nineteenth century progressed more were taken over. The brewers also began to build their own, often splendid, houses in the larger towns and cities. 'Fewer and better' became a phrase used by the brewers, and this pattern was repeated throughout Britain. Today, of about seventy-five thousand full on-licences in the United Kingdom, about fifty thousand are owned by the breweries and twenty-five thousand are free houses.

It was also during the nineteenth century that organised opposition to beer reached its peak. In 1839 public houses in London were forced to close from midnight on Saturday until midday on Sunday – a seemingly odd regulation to those of us used to twentieth-century licensing laws, but one which had a practical use at the time. It was, as the authorities pointed out, in those twelve hours that most drunkenness occurred! All-day Sunday closing was introduced in Scotland in 1853, in Ireland and Wales in 1881. Fortunately, in spite of considerable pressure, England held out.

ABOVE: *Burton upon Trent became famous for the quality of its pale ales and bitter beers, but Burton brewers were restricted in their distribution until the railways provided cheap and effective transport to all parts of Britain. Bass developed the biggest private railway system in Europe, designing its own engines and maintaining hundreds of wagons and trucks for distributing its beers.*

BELOW: *Today many breweries have revived the use of Shire horses both for local deliveries and for show. Here, a pair of Young's horses, which are a regular sight on London's streets, deliver beer to the Green Man in Putney.*

In America individual states were introducing bans on the public sale of drink, and temperance movements in Britain began to try to get similar legislation on to the statute books. The Church of England Temperance League was established in 1862 to coordinate the fight. However, the most famous organisation was the Band of Hope founded by Jabez Tunnicliff.

By the end of the nineteenth century, however, beer production in Great Britain was around forty million barrels, brewers had become more specialised, the techniques of brewing had developed as never before, but there was a cloud on the horizon: a recession was looming.

ABOVE: *Steam wagons were used by the brewers for deliveries, but their period of use was short: many breweries kept their horses and the advent of the petrol engine effectively killed the use of steam. This Foden was one of three used by Hall and Woodhouse of Blandford Forum from 1912 to 1922.*
BELOW: *Brewery drivers have always been proud of their vehicles. This impressive line of Bentley and Shaw's Leylands dates from the 1930s. They are standing in front of the company's maltings: the pitched roofs with the capped 'chimneys' are the kilns where the malt is dried and the long building on the left contains the open malt floors. Bentley and Shaw's brewery was in Huddersfield.*

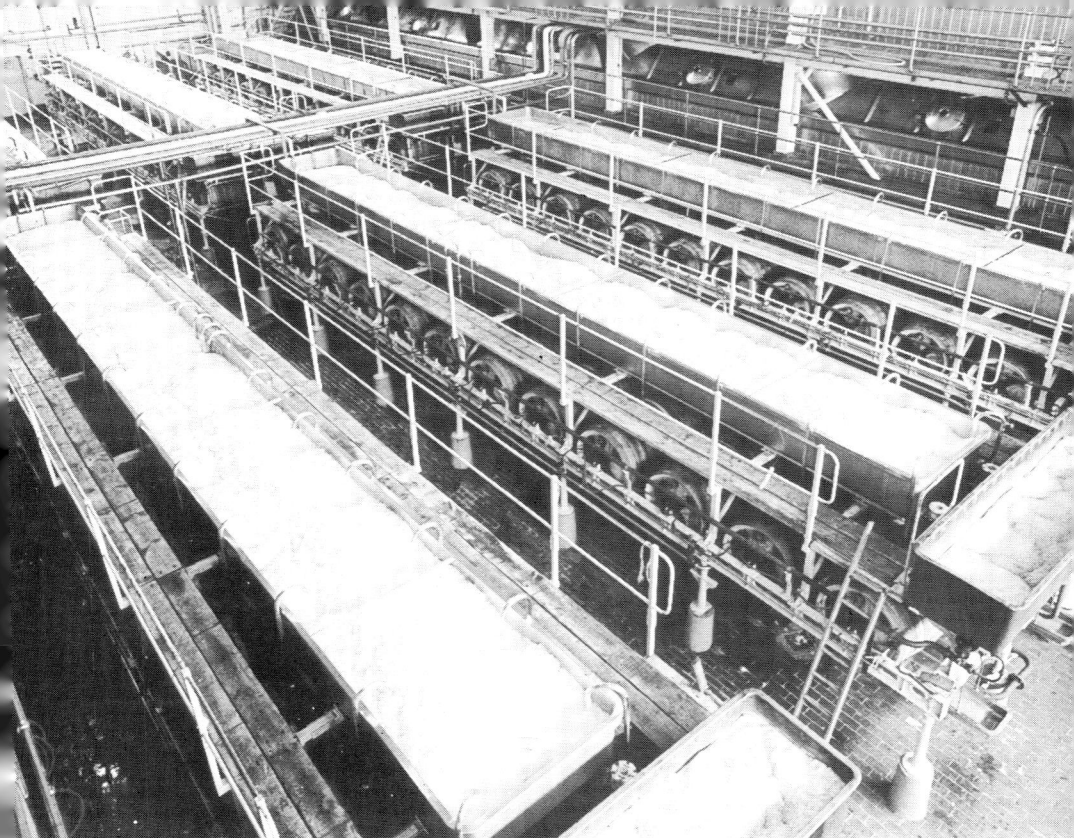

Because of the resurgence in interest in traditional cask beer, one brewer, Marston's of Burton upon Trent, has retained its famous nineteenth-century Burton union system of fermentation. The fermenting beer is run into giant oak casks, called unions, where the fermentation process is completed. The yeast rises to the top of the casks and then through the swan-neck pipes into the top trough, where it can be collected. The system, once used by all the main Burton brewers, but now exclusive to Marston's, gives their beers a unique flavour.

THE TWENTIETH CENTURY

Economic depression substantially reduced demand for beer and many famous companies either went out of business or joined forces with their neighbours to try to escape financial difficulties. The First World War created even more problems, not the least of which was the Defence of the Realm Act, which established restricted licensing hours, supposedly only for the duration of the war! Introduced to control the times at which alcohol was sold in areas where munitions were made, it was later extended to cover the whole of Britain by Lloyd George, who is reputed to have said: 'Drink is causing more damage in the war than all the German submarines put together.'

After the war attempts were made to put brewing back on to a sound basis, but the depression years limited demand and even more breweries disappeared or, once again, formed alliances to try to solve the problems of insufficient places to sell their beer and breweries which were too big and employed large numbers of workers.

New beers were introduced to try to encourage a new demand. Lager, which had first been brewed and sold (albeit unsuccessfully) in Britain in the 1880s, was tried again but failed to make much impact until after the Second World War. Bottled beer assumed a new importance as the brewers began to look for new markets.

During the 1930s a radical change began to take place. Brewers began to rely more on advertising to sell their beers and also to develop new ranges of products to attract customers. The first canned beers, for example, were introduced by the Felinfoel Brewery in Wales in 1935, opening up a new market for beer which could be easily and conveniently drunk at home. The new beers and their advertising began to attract new customers, but the Second World War curtailed production. All the beer that could be brewed could be sold, so the necessity for aggressive advertising and marketing was reduced.

After the war brewing was still beset by problems. Britain was becoming more of a consumer society, and people were demanding more for their money. In beer and pubs, variety was the key. The English pub is unique, but two world wars and a long period of recession had left many pubs in poor condition. In the 1950s it was realised that pubs should be made more attractive, and in the 1960s and 1970s efforts were made to provide variety in pub types, more than matching the great nineteenth-century developments. Brewers wished to attract the uncommitted drinker to their particular houses, and the beer was only one of the reasons for choosing a favourite pub. Pubs became restaurants, discotheques, the headquarters of the local football team, meeting places, the bases for many local societies, the village hall, the place for the best food at low prices – in short far more than a place just to have a drink, the traditional picture of the pub.

However, the majority of the British beer business was being increasingly controlled by a few companies. The choice of beers was seriously restricted if one company owned all the pubs in a particular area; also most of the supposed 'free houses' were actually tied to one brewer who, through large loans, could control an even greater share of the market by insisting that only its beers could be sold. The smaller or local brewers were therefore effectively frozen out of the national marketplace.

By the early 1990s the government was under considerable pressure to investigate the stranglehold of the big brewers on the industry. In 1992 the Monopolies and Mergers Commission recommended major changes to the industry. After much wrangling, the most important of these were embodied in a series of DTI Orders, the main one being a restriction on the number of pubs that a brewer could own, if it wanted to be both a brewer *and* a retailer.

Some of the larger brewers sold off their breweries; others began massive disposal programmes and, almost overnight, hundreds of pubs came on to the market – one major brewer had to dispose of more than three thousand properties. However, the attractiveness of the concept was overtaken by the reality of making it work. Many of the smaller brewers, who the government said would benefit from the change in being able to supply their beers to a bigger 'free trade', simply sold off or closed down their breweries and invested the capital in buying pubs, usually from the major brewers' enforced sales. The beers would come from a variety of sources, often from major brewers who now had a surplus capacity.

Sadly, the glut of pubs on the market resulted in many closures. But the expanded estates of the independent brewers did provide more choice in many areas and over fifty new pub groups were established, largely by entrepreneurial companies. Because they had no traditional links with brewing, most brought innovative ideas into the business with new styles of pub and the improvement of tired and neglected premises. Very small brewing companies and pubs brewing their own beers have also seen a revival. However, their problem remains getting sufficient distribution to make their activities financially viable.

There has also been a resurgence of interest in traditional British draught or cask beers – 'real ales' – served in traditional surroundings. The use of top-fermenting yeast in the brewing process gives them their best flavour when served at 13°C (56°F). Visitors to Britain look on aghast, but chilling destroys the flavour, as any beer drinker knows. There is also more interest in premium bottled beers and lagers, many of which are imported, often from those parts of the world with brewing traditions as strong as Britain's: Germany, Belgium, the Czech Republic, the USA and even Japan. British brewers are also looking back to their traditions and reviving or recreating interesting regional bottled beers with strong and distinctive flavours. The drinker can now try such exotic brews as Strong Suffolk, Spitfire, Bishop's Finger, Thomas Hardy's Ale, Jacobite Ale, Headcracker and even several examples of what was once an eighteenth-century ale, the dark, strong and full-flavoured porter.

But probably the greatest change has been in pub food, now served in a variety of choices of hot and cold dishes in purpose-built areas or even full restaurants. The introduction of all-day opening means that many pubs now offer breakfast, morning coffee and afternoon tea in addition to lunch and dinner. Pub food in the United Kingdom, worth more than £3.5 billion in 1996, is the biggest single sector of the eating-out market. It is larger than the combined turnover of all the fish and chip shops, hamburger bars and pizza bars, with an estimated 1,300,000 pub meals served every day.

Today there are some 83,000 full on licences in the United Kingdom, of which 25,000 are free houses, 27,000 are tenancies and 13,000 are managed. Over 622,000 people are employed in pubs and clubs. The brewing business is also a major contributor to the economy, with the tax revenue generated by pubs and clubs in excess of £9,200 million.

Since 1980 there has been a major shift in leisure habits. Working time has been decreased, leisure time increased. Home entertainment has experienced a major boom. Yet there is still a vital need for people to have somewhere to meet, for social contact and a fresh environment. For centuries, the English pub has provided that place and it will continue to do so.

In 1617, Fynes Morrison wrote: 'The world affords not such great Innes that England hath, either for food and cheap entertainment after the guest's own pleasure, or for humble attendance.' He would certainly have approved of brewing and pubs today.

FURTHER READING

Barber, Norman. *A Century of British Breweries*. Brewery History Society, 1994.
Boston, Richard. *Beer and Skittles*. William Collins, 1976.
Corran, H.S. *A History of Brewing*. David & Charles, 1976.
Davis, Ben. *The Traditional English Pub*. Architectural Press, 1981.
Dunkling, Leslie, and Wright, Gordon. *A Dictionary of Pub Names*. Routledge & Kegan Paul, 1987.
Evans, Jeff. *Camra Good Beer Guide*. Camra Books.
Finn, Timothy. *The Pub Games of England*. Oleander Press, 1981.
Gourish, Terry, and Wilson, Richard. *The Brewing Industry in England 1830-1980*. Cambridge University Press, 1994.
Jackson, Michael. *The World Guide to Beer*. Mitchell Beasley, 1994.
Jackson, Michael. *Pocket Beer Book*. Mitchell Beasley, 1995.
Mathias, Peter. *The Brewing Industry in England 1700-1830*. Cambridge University Press.
Peaty, Ian. *Brewery Railways*. David & Charles, 1986.
Protz, Roger. *The Real Ale Drinker's Almanac*. Lochar Publishing, 1990.

CAMRA (Campaign for Real Ale Ltd), 230 Hatfield Road, St Albans, Hertfordshire AL1 4LW (telephone: 01727 867201), publishes a newspaper, *What's Brewing*, for members and produces a range of books, including the *Good Beer Guide*.

GLOSSARY

Alcohol content: the percentage by volume of alcohol in beer.

Ale: historically, an unhopped, fermented malt drink; nowadays ale is any beer produced by fermentation at water temperatures of around 65-70°F (18-21°C).

Alpha acid: the main resin contained in hops which is converted to bittering substances during boiling of the wort.

Back: traditional brewing term for a vessel, e.g. underback, hopback.

Barley wine: a strong ale.

Barm: liquid yeast recovered after fermentation for reuse.

Barrel: a 36 gallon (163.656 litre) cask.

Batch fermentation: the traditional method of allowing wort to ferment in individual batches.

Beer: historically, a fermented beverage brewed with hops.

Beer engine: suction pump operated by hand or by electric motor, to deliver beer to the bar counter.

Bitter: pale or amber ale, usually well hopped for a bitter flavour.

Bottom fermentation: traditionally, method of fermentation to produce lager. During fermentation, the used yeast sinks to the bottom of the fermenting vessel.

Brewery-conditioned beer: beer which has been brought to perfect condition in the brewery prior to filtration, instead of conditioning in cask or bottle.

Bright beer: beer which has been filtered to make it brilliant to the eye.

Brown ale: traditionally a bottled mild ale.

Burtonisation: the treatment of water with minerals to make it similar to water found in Burton's wells.

Caramel: sugar darkened by heat treatment and used for adjusting the colour of beer.

Carbon dioxide: a gas produced during fermentation.

Cask: general term for all draught beer containers, whatever the size, now almost always made of metal.

Cask-conditioned beer: beer which conditions in the cask instead of in conditioning tanks at the brewery.

Casting: the art of emptying a brewery vessel, e.g. the copper.

Centrifuge: a machine designed to separate excess yeast from beer.

Chilled and filtered beer: beer which is conditioned in the brewery and then chilled, so that solids such as yeast sink to the bottom. It is then filtered to remove any remaining particles.

Conditioning: maturation of beer after it leaves the fermenter.

Conical-bottomed fermenter: cylindrical fermenting vessel, erected vertically. It has a conical base for more convenient yeast removal.

Cooper: one who makes and repairs wooden casks.

Copper (or kettle): copper or stainless steel vessel in which wort is boiled with hops to give beer its bitter flavour.

Diastase: enzymes in malt which cause the conversion of starch to simple sugars in the mash tun.

Draught: term describing beer which is drawn through a tap or pump to the bar.

Dry hopping: the addition of hops to beer during maturation to improve flavour and aroma.

Enzymes: agents which cause changes from one substance to another, present in all living things. In the mashing process they convert carbohydrates into brewing sugars.

Excise duty: the tax on beer.

Fermentation: the process in which yeast converts wort into beer.

Fermenting vessel: vessel where fermentation takes place and in which the volume and specific gravity of wort is declared to the Customs and Excise so that duty can be levied.

Filtration: removal of solid particles from beer using a filter unit.

Fined: beer which has been clarified by the addition of isinglass finings.

Finings: colloidal substance, produced from isinglass, which is added to beer to assist in its clarification.

Firkin: a 9 gallon (40.914 litre) cask.

Green malt: the name given to malt before kilning.

Grist: crushed malt, ready for mashing.

Grist case: grist storage hopper, situated above the mash tun.
Hogshead: a 54 gallon (245.484 litre) cask.
Hop: a perennial climbing plant; its 'cones' or flowers give beer a bitter flavour and aroma.
Hopback: vessel designed to remove hops from boiled wort.
Hop extract: bittering substance produced from hops and concentrated into syrup.
Hop pellets: natural hop powder which has been compressed.
Hop pocket: traditional sack used to deliver hops to the brewery, usually holding 176 pounds (80 kg).
Hop powder: natural hops finely shredded.
Hot liquor tanks: vessels where water is heated before mashing.
Isinglass: semi-transparent colloidal substance obtained from swim bladders of sturgeon; the raw material of finings.
Keg: sealed container usually for chilled and filtered beers.
Kilderkin: an 18 gallon (81.828 litre) cask.
Kilning: the drying and curing of green malt by heat treatment.
Lager: a straw-coloured fermented malt drink made from lightly kilned malts and fermented at 45-55°F (7-13°C).
Lauter tun: a filtration vessel used in modern mashing techniques.
Light ale: a particular type of bottled ale.
Liquor: the brewing term for water.
Malt: barley which has been steeped in water, allowed to germinate and then heated in a kiln to halt germination.
Malt extract: sugars extracted from malt and concentrated by evaporation.
Malt mill: machine which crushes malt into grist.
Mashing: mixing together of grist and hot water at precise temperatures to form malt sugars, which the yeast will eventually ferment.
Mash tun: vessel in which mashing takes place and in which wort is separated from the grains (spent).
Maturation: the storage of beer for a period during which its quality improves.
Oast house: kiln where hops are dried.
Original gravity: the specific gravity of wort before fermentation. Excise duty is paid on this gravity. The higher the gravity the higher the tax paid.
Pale ale: beer brewed from pale malt, either on draught or in bottle. East India pale ale, or IPA, got its name because it was originally brewed for the Indian market.
Pin: a $4^{1}/_{2}$ gallon (20.457 litre) cask.
Pitching: adding yeast to wort in the fermenting vessel.
Priming sugar: sugar solution added to beer during conditioning.
Racking: filling casks with beer.
Secondary fermentation: process in which beer continues to ferment and mature slowly in cask or bottle.
Shive: wooden bung in the top of a cask with a central core. A wooden peg (spile) is driven through the centre and allows carbon dioxide to escape.
Skimming: removing yeast from the top of beer as it ferments.
Sparging: spraying hot water over mash in the mash tun through a rotating arm to ensure complete extraction of malt sugars.
Sparkler: adjustable nozzle which regulates the flow of beer from a beer engine; used to adjust the amount of 'head' on a pint.
Specific gravity: a measure of density of liquids.
Steeping: the exposure of barley to moisture to commence germination – the first stage of the malting process.
Stillage: a brick, wooden or metal structure which supports casks of beer in the pub cellar.
Sweet wort: unboiled wort before the addition of hops.
Top fermentation: traditional method of fermenting wort in which yeast rises to the top of the beer during the process.
Trub: malt protein material which coagulates during boiling. It is removed before fermentation.
Underback: vessel, usually below the mash tun, which collects wort during sparging.
Venting: the use of hard and soft spiles to control carbon dioxide escaping from a cask.
Wort: unfermented beer.
Yeast: single-celled micro-organism which brings about fermentation.
Yeast press: a press for squeezing beer from surplus yeast.

PLACES TO VISIT

To avoid a wasted journey, intending visitors should telephone ahead for times of opening.

Bass Museum, Horninglow Street, Burton upon Trent, Staffordshire DE14 1YQ. Telephone: 01283 511000 or 542031.

Chiltern Brewery, Nash Lee Road, Terrick, Aylesbury, Buckinghamshire HP17 0TQ. Telephone: 01296 613647.

Jennings Brewery Tour, The Castle Brewery, Cockermouth, Cumbria CA13 9NE. Telephone: 01900 823214.

Stamford Steam Brewery Museum, All Saints Street, Stamford, Lincolnshire PE9 2PA. Telephone: 01780 52186.

Tetley's Brewery Wharf, The Waterfront, Leeds, West Yorkshire LS1 1QG. Telephone: 0113-272 0666.

Tuckers Maltings, Teign Road, Newton Abbot, Devon TQ12 4AA. Telephone: 01626 334734.

Weymouth Museum, Timewalk and Brewery Museum, Brewers' Quay, Hope Square, Weymouth, Dorset DT4 8TR. Telephone: 01305 777622.

Whitbread Hop Farm, Beltring, Paddock Wood, near Tonbridge, Kent TN12 6PY. Telephone: 01622 872068.

Many breweries organise brewery visits and tours, primarily for groups and usually at a month or more's notice. A list is available from the Brewers and Licensed Retailers Association, 42 Portman Square, London W1H 0BB (telephone: 0171-486 4831).

The following breweries have visitor centres, open regularly, especially during the tourist season, with displays about brewing and products for sale.

The Black Sheep Brewery, Wellgarth, Masham, North Yorkshire HG4 4EN. Telephone: 01765 689227.

Elgood & Sons Ltd, North Brink Brewery, Wisbech, Cambridgeshire PE13 1LN. Telephone: 01945 583160.

Flagship Brewery, Unit 2, Building 64, The Historic Dockyard, Chatham, Kent ME4 4TE. Telephone: 01634 832828.

Hogs Back Brewery, Manor Farm, The Street, Tongham, Surrey GU10 1DE. Telephone: 01252 783000.

St Austell Brewery, 63 Trevarthian Road, St Austell, Cornwall PL25 4BY. Telephone: 01726 74444.

T. & R. Theakston Ltd, Wellgarth, Masham, North Yorkshire HG4 4DX. Telephone: 01765 689544.

Tollemache and Cobbold Brewery, Cliff Road, Ipswich, Suffolk IP3 0AZ. Telephone: 01473 231723.

Traquair House Brewery, Innerleithen, Peeblesshire, Scotland EH44 6PW. Telephone: 01896 830323.

Working reconstructions of nineteenth-century public houses can be found at Beamish, The North of England Open Air Museum, County Durham DH9 0RG (telephone: 01207 231811) and at the Black Country Museum, Tipton Road, Dudley, West Midlands DY1 4SQ (telephone: 0121-557 9643).